T0117531

SOUTHERN RAILWAY.

ONE BICYCLE (accompanied by Passenger)

(E.)   Victoria to   (E.)

2/4   ANY STATION
NOT EXCEEDING
75 MILES

This ticket is available for one journey only
and must be given up at destination Station.

FOR CONDITIONS SEE BACK.

2183   2183

The Bicycle

A Miscellany
on Two Wheels

Peter Ashley

ACC ART BOOKS

Dedicated to everybody who rides
a bicycle, keeps one in the shed
or just remembers the exhilaration
and sense of freedom when the
stabilisers came off

Introduction

I graduated from three wheels to two when my father brought home an anonymous fixed-wheel bicycle from Friswell's in Applegate Street, Leicester. I've lost the photograph of me standing very proudly with it, but it may have been scribbled on after I'd ridden it into the lamppost opposite our house and slid down the crossbar into the handlebar pillar with devastating and painful effect. But soon it was the memorable Christmas morning when I was led down the lane to Mr. Dalby's bungalow where a purple and green Elswick was hiding in his greenhouse.

And so my hitherto limited horizons widened. The interminable journeys to and from school with mates, being shouted at by prefects for not wearing our caps, being ignored by girls as we whistled at them on their Triumph Pink Witches. The slightly subversive air of the bike shed ("Don't stay too long in there," warned my mother enigmatically), the Elswicks jostling with Phillips, Dawes and Carlton bicycles. On Saturdays and holidays we rode along the towpath of the Grand Union Canal, annoying fishermen, and undertook expeditions to remote country churches with bananas and Vimto in our saddlebags. Until suddenly I appeared to be seventeen, and a rusting Ford Thames van was quietly leaking oil onto the driveway.

The bicycle has its origins in the impossibly heavy wood and metal velocipedes at the beginning of the nineteenth century, which would appear from contemporary engravings to be slightly slower than walking – this was the motive power for them anyway. How it became to be mechanically propelled appears to be claimed by any number of inventors, usually French, but I plumb for Scottish blacksmith Kirkpatrick Macmillan in 1830. Then James Starley brought us the 'ordinary' bicycle in 1870, quickly dubbed the 'penny farthing' owing to the massive disparity in size between the front and back wheels. This was the first machine to be called a 'bicycle' but it was his nephew John Kemp Starley who welded the frame into the triangular shape we now recognise and put the still standard 26" wheels on it. He called it the Rover, which summed up all the possibilities as nicely as it did for his first car in 1904.

Have we ever become such a single entity with a machine? My mother, a Baptist minister's daughter, flying on her cycle down The Avenue, a steep tree-lined walk in Wellingborough, doing her circus act of standing up on the saddle with her arms outstretched until having an umbrella waved at her by a member of her father's congregation. I took for granted that the wretched lamppost I embraced was lit every night and extinguished every morning by a lamplighter from the local gasworks arriving on a bicycle with a ladder sticking out front and back. The policeman who quizzed me about a broken window leaning his obligatory bike against our garden fence under billowing hydrangeas. Arthur Seaton sullenly riding over Nottingham cobbles to his Raleigh factory lathe in *Saturday Night and Sunday Morning*, the baker's boy in Ridley Scott's Hovis commercial labouring up Gold Hill in Shaftesbury and joyously flying back down again. And possibly the most angst-ridden bicycle in fiction – Leo Colston's carefully wrapped birthday present in L.P. Hartley's *The Go-Between*.

It was the early 1980s before I ventured out on a bicycle again, a large and exceptionally sturdy black Raleigh as ridden by my broken-window nemesis. Curiously I fell off it after cycling around the interior of The Clifton pub in St. John's Wood one lunchtime. There are still bikes that appeal to me now, but I know that if I bought one it would be seen as an ambient prop and talking point rather than being ridden for muscular long distances. So I content myself with the *thought* of cycling, more parcel-laden postman and whistling butcher's delivery boy than a mountain biker on The Fens; more country vicar with communion wine in the wicker basket and a squeaking cotter pin than being submerged in a peloton of would-be Touristes de France shouting through villages.

As you will see, the art of balancing on two wheels will be viewed here through the lens of my doubtful prejudices, where a yellow Dunlop Repair Outfit tin is just as appealing as a Boris bike outside an Underground station, and the joke of sitting on a bicycle not knowing that the saddle's been removed is just as childishly funny sixty years on. But cycling really is healthy, life-affirming and great fun and all you need is a well-oiled bicycle with the tyres pumped up. And a saddle.

GAMAGES
"POPULAR" CYCLE

is known in every village in the country, it has led the field for over **40** years and is still the world's best value――――

Gamage's latest Cycling Catalogue sent free on request

Carriage paid to nearest Railway Station England and Wales. Scotland slightly extra

JUST COMPARE THIS SPECIFICATION

FRAME & FORKS—Built with genuine Reynolds Weldless Steel Tubing.

SIZES—22″, 24″ and 26″. Low built with 26″ wheels. Light Roadsters, with sloping top tube and "North Road" Handlebars, 22″, 24″. Boys' and Girls' 18″, 20″, 24″ wheels to 18″ models.

WHEELS—Built with Phillips' Oil-Retaining Hubs, genuine Dunlop Rims, Black Enamelled Spokes.

HANDLEBARS—Phillips' "Philco" with two powerful Roller Lever Brakes.

FREEWHEEL—Genuine Villiers.

CHAIN—Coventry.

CHAINWHEEL & CRANKS—Williams.

PEDALS—Phillips.

SADDLE—Brooks' "Legion" with three Coil Springs.

TYRES—Genuine Dunlop Roadster, 26″ × 1¼″ W.O.

FINISH—The Frame is Costettised, making it absolutely rustproof. This is followed by a special undercoating which, in turn, is followed by three coats of lustrous Black and lined Red and Gold. All bright parts are heavily Chromium plated.

Free Accessories

PUMP, PUMP-CLIPS, TOOLS, TOOL-BAG AND REFLECTOR

CASH PRICE
Gents' and Boys' Model .. **£3-10-0**
or 12 Monthly Payments of 6/6
LADIES' AND GIRLS' MODELS .. **£3-17-6**
or 12 Monthly Payments of 7/-
3-Speed Gear 20/- extra

GAMAGES, HOLBORN, LONDON, E.C.1   'Phone : Holborn 8484.   Also at 107 Cheapside, E.C.2

above **Symonds' Standard, 1901. '...swaged piano-wire spokes, climatically tested rims... a marvel of good value'**

opposite **Gamages Popular, c1949. 'Carriage paid to nearest Railway Station England and Wales. Scotland slightly extra'**

A LABOUR OF LOVE.

Amorous Cycling Lessons

above **'A Labour of Love'
posted in Forest Gate,
London in 1908**

opposite **Only the first lesson
and he's already taking off his
bowler**

After the first lesson.

An Edwardian advertisement
illustration showing that as
a bicycle is so easy to ride
a girl will have no problem
riding no-handed in order to
pull her gloves on. Or off, and
at speed

'...the greater number of my fellow-travellers were cyclists.....In one band seven or eight lean young chaps in dark clothes bent over their handle-bars, talking in jerks as they laboured, all stopping together at any call for a drink or to mend a puncture. They swore furiously...'

In Pursuit of Spring
Edward Thomas 1914

ARE **YOU** FOND OF CYCLING?
IF SO
WHY NOT CYCLE
FOR THE KING?
RECRUITS WANTED
By the S. Midland Divisional Cyclist Company
(Must be 19, and willing to serve abroad).
CYCLES PROVIDED.   Uniform and Clothing issued
on enlistment.

Application in person or by letter to
Cyclists, The Barracks, Gloucester.
BAD TEETH NO BAR.

The Bicycle at War 1

above **Recruiting poster for the South Midland Divisional Cyclist Company, active in both world wars**

opposite **A 1900 print of the Lancashire Fusiliers, cycling into the Boer War in South Africa**

FOR BUSY MEN.

Constables Ride Them.

Barristers Enjoy Them.

FOR THE NURSES.

JOY FOR CLERKS.

CommercialsWant Them.

The great democratisation of bicycles, from a c1915 advertisement

RIDE A

RALEIGH

THE ALL-STEEL BICYCLE

Raleigh Vitreous Enamel Signs

above **Raleigh used the tag line 'The All-Steel Bicycle' to put one over on competitors who still used cast iron for critical parts**

opposite **Turnastone, Herefordshire. Note the additional helpful mileage sign and the desirable Castrol enamel**

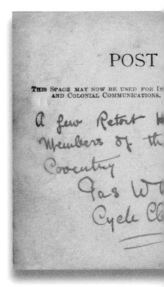

POST

THIS SPACE MAY NOW BE USED FOR IN
AND COLONIAL COMMUNICATIONS.

a few Retort W
Members of the
Coventry
Gas W
Cycle Cl

For industrial workers who toiled for long hours in filthy and inhospitable conditions, fresh air was a longed-for commodity. Many big factories and plants like the Coventry Gas Works formed very popular cycling clubs, and here are a few members employed in producing gas by heating coal in retorts

This is a cast iron plaque for the Cyclists' Touring Club. Founded in 1878, it was the first touring club in the world, and the winged wheel plaque was used on approved hotels and restaurants where members would find good service and attention. In 2016 it was renamed Cycling UK

CYCLING

Road Book & Guide
to the
ENVIRONS OF LONDON

With Coloured Map
(2½ Miles to Inch)

1/-NET

By A·C·ARMSTRONG
and
HARRY·R·G·INGLIS

VOL. 1 - South of the Thames

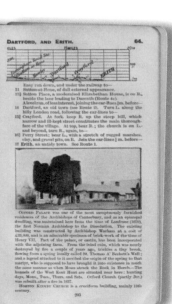

opposite **One of *Cycling* magazine's volumes (c1900) giving cyclists routes to follow, here those around London to the south of the Thames**

top **Choose your numbered route and then *left* follow the instructions, which give brief notes of what to see from the saddle: 'Erith, an untidy town', with gradient maps and notes: 'Dangerous crossing here over tram-lines, and a short dip under the railway'**

Novelist and poet Thomas Hardy with his Rover Cob outside Turnworth Rectory in Dorset around 1899. In a letter he thought that cycling was 'a delightful escape, of a temporary kind, from the cares which crowd in upon one as one grows older'

There has been a cycle shop on this site in Market Harborough since 1898. Here are the premises in the 1930s, now the home of George Halls Cycle Centre. I particularly like the tyres hanging up like pheasants and rabbits outside a game dealer's

PLAYER'S CIGARETTES

"INVINCIBLE" BICYCLE

Player's Cigarette Card
No.12 'Invincible' Bicycle

The penny farthing must
have been very impractical
to ride, perhaps needing a
friend to get you going. Here
the 59″ 'Invincible' is
ridden by H.L. Cortis in 1882

PLAYER'S CIGARETTES

COMPANION SAFETY BICYCLE

Player's Cigarette Card
No.20 Companion Safety
Bicycle

What a very companionable
way to cycle in 1896.
Although those dual handle-
bars and only one steering
front wheel might cause a
problem

above **Rehearsal for the American Jubilee show at the New York World's Fair 1939–40 in Flushing Meadows, Queens**

opposite **Zwei Mädchen mit Fahrrädern, c1900. A classic studio photograph of two German girls proud of their bicycles**

CYCLING

Touring Guides

2'6
NET

No 4   SOUTH-WEST ENGLAND

Cycling Touring Guides
written by Harold Briercliffe
and published by Temple
Press in the late 1940s

The insouciance of being
able to smoke and cycle at
the same time. Bicycle
Cigarettes from Wills

The insouciance of card
playing and cycling. But did
'Santé et Plaisir' (Health and
Pleasure) refer to cycling?

above **A Joseph Lucas bicycle bell, c1914. This particular bell has a very loud resounding ring, which must have sent unwary pedestrians leaping for the side of the road**

right **Bee Brand revolving bell from Shanghai, China. Perhaps from one of Katie Melua's 'nine million bicycles in Beijing'**

opposite **Lucas lighting for those nocturnal cycling activities**

LUCAS "King of the Road"

CYCLE LIGHTING EQUIPMENT
AND ACCESSORIES

For Quality
and
Dependability

JOSEPH LUCAS LTD · BIRMINGHAM

above **Elephant on a bicycle**
Harry B. Neilson 1861–1941

opposite **Squirrel on a bicycle**
Peter Ashley, 1977

The Bicycle at War 2

above **German World War II bicycle with a 'panzerfaust', a preloaded recoilless single-shot anti-tank weapon with a range of 60 metres**

opposite **Home Guard improvisation on an exercise in North Wales, 1st November 1943**

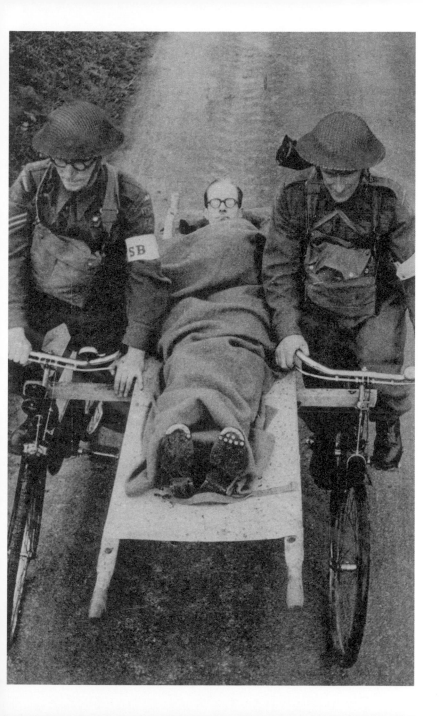

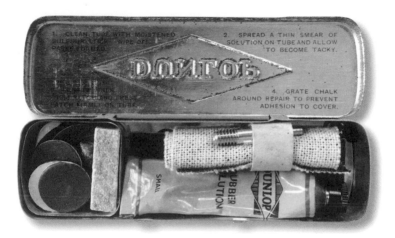

First essential for a lengthy
bike ride: the puncture repair
outfit. Second essential:
a companion rider who
knows what to do with it all

Seats of Learning

above **Kingston Road, Oxford**
(Photo: Nick Wright)

opposite **Trinity Lane,
Cambridge**

PLEASE DO NOT LEAN CYCLES
AGAINST THIS WALL

The Pashley Guv'nor, a revival of their 1930s Path Racer. Now Britain's longest established cycle company, founded by William 'Rath' Pashley in 1926. When I first saw the name I was of course convinced they were made just for me. The Guv'nor certainly was

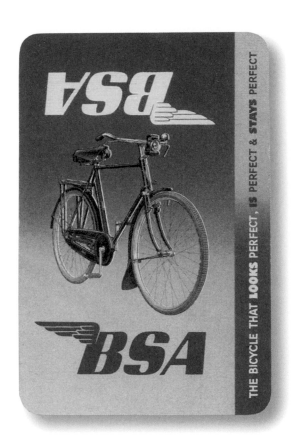

BSA was the Birmingham Small Arms Company, which provided military and sporting firearms from 1863 to 1973. And bicycles that – we were assured – both looked perfect and stayed perfect

The United States Playing Card Company cashed in on the new cycling craze in 1885 by introducing Bicycle cards. 'Rider back' refers to Cupid on the card-reverse riding a bike; 'air-cushion finish' to a plastic coating

above **Cycling** was first
published in 1891; this is the
New Year issue for 1956, with
this intrepid cyclist *right* about
to start new adventures

opposite **Frank Patterson
provided illustrations for the
journal from 1893 until his
death in 1952, serving under
six editors. Here is his atmos-
pheric rendering of a snowy
Sussex, his home county**

ALFRISTON

Turnabout Street
AMBERLEY

The catslide roof of
WESTMESTON CHURCH

EXPERIENCE IN A DRIFT OFF THE BEATEN TRACK
NEAR POYNINGS

SUSSEX
UNDER SNOW.

A BIT OF BILLINGSHURST IN CLOSE TOUCH
WITH "CYCLING"

THE G.P.O. BICYCLE — POOR CHAP !

Bicycles at Work 1

Exercising hounds up above the Welland Valley in Leicestershire with two market-coated kennelmen on bicycles, one to drive them forward *above* **and the other** *opposite* **to lead and clear the way**

Painted Bicycles 1

Village Street
Eric Ravilious (1936)

Ravilious never learned to
drive, and so cycled in all
weathers everywhere to
paint. The couple here are
thought to be Bennett Smith,
owner of the hardware store
in Great Bardfield, and his
wife, setting out for a cycle
ride in the Essex countryside

Cycling Home
Simon Palmer (1996)

Palmer's meticulous and enigmatic paintings in ink, watercolour and gouache often find inspiration in the tree-lined lanes, stone walls and farm buildings of North Yorkshire
(© 1996 Simon Palmer)

above **A big box of matches by Rex London. Excellent for lighting a wayside Primus stove**

top **Matches made in Sivakasi (in the Indian state of Tamil Nadu), where the main industries are firecrackers and printing. And matches**

An uncut sheet of matchbox labels from the Czech Republic. I like the ornamental frock guard on the bicycle bottom right and the chain guard on the one above

Hallaton, Leicestershire

above **The ghost bicycle that appeared in the car park of The Fox Inn**

opposite **A highly original entry in the village's scarecrow festival, Chris Froome by Chris Clarke**

Posting Fun 1

'Pleasing Ma-In Law', c1907.
The recipient was told on the
reverse: 'Your old fatheaded
uncle went away with your
ticket by mistake'.

"OUR JOE'S HAD A TERRIBLE ACCIDENT—HE TOOK A FLYING LEAP ON TO HIS BIKE, AND DIDN'T NOTICE SOMEONE HAD PINCHED HIS SADDLE!"

Posting Fun 2

What a change in tone fifty years later. The unmistakable style of a Bamforth postcard with a Ben Fitzpatrick illustration. Poor Joe

(© Bamforth & Co.)

Raleigh Cycles have featured on the British scene since they started in a bicycle shop in Raleigh Street, Nottingham in 1885

above **The Raleigh badge. The heron's head is from co-founder Frank Bowden's family crest**

opposite **Detail of a 1970s Raleigh advertisement, very likely in the midst of a 'Buy British' campaign**

John Boultbee Brooks started making leather harness and tackle for horses in 1866. Trying out the new 'bicycle' in 1878 he found the saddle so painful to sit on he decided to make his own. And so was born what many believe are the ultimate saddles, none more so than the one shown here, the B17, which was launched in 1888 and remains Brooks' bestselling saddle

In 1891 brothers Édouard and André Michelin took out a patent on a cycle tyre that didn't have to be glued to a wheel rim. Their removable pneumatic tyres were used by Charles Terront to win the Paris-Brest-Paris race in the same year

above Bibendum enjoying his new Michelin tyres

opposite The original cigar-chomping Bibendum rides across a window at Michelin House, Fulham Road, London

The Bicycle as Marketing Tool

above **What started out as the idea of keeping your saddle dry by covering it with a plastic carrier bag was soon snatched up as a unique media opportunity. These are in the City of Bicycles – Cambridge**

left **Transport for London cycle roundel**

opposite **Boris bikes outside Warwick Avenue Underground Station, London**

Bicycles at Work 2

Most rural railway stations
would keep a bicycle for
delivering parcels locally.
Here is the Southern Railway
bicycle at Horsted Keynes
station on the Bluebell Line
in East Sussex

The Delibike from Pashley. It's very heartwarming that the delivery bicycle is still being made. Essential for butchers' and grocers' boys and girls, whistling along in striped aprons and boaters like this *left* illustration by the author

above **Victorian Penfold pillar box as bicycle prop, King's College, Cambridge**

opposite **Pembroke College, Cambridge. The wicker basket is an essential accessory for Nietzsche and a Penguin *Brideshead Revisited***

PC Beveridge teaches
cycling proficiency at Millfield
County Primary School in
Braunstone, Leicester in
1972. Equally important was
the mechanics of how a
bicycle worked and should
be looked after. 'Make sure
that chain's kept tight!'
(Photo: Leicester Mercury)

above **The National Cycling Proficiency Scheme was introduced by the government in 1958. This is its hard-earned badge. Now it's called the National Standard for Cycle Training, which seems to be something else entirely. Fortunately it's also called 'Bikeability'**

top **National Road Safety Campaign badge**

above **Phillips Bicycles were made in Smethwick near Birmingham, and as a British producer were once second only to Raleigh, who took them over in the 1980s**

left **Phillips for Christmas '54**

opposite **Detail from a Coronation advertisement from June 1953. Buy a Phillips bike and pigeons will feed from your hands**

Painted Bicycles 2

**Code 1057: cycle lane, track
or route, schedule 6 (road
markings), Traffic Signs
Regulations 1982**

above **Kettering Road, Market
Harborough, Leicestershire**

opposite **On the redundant
runway of what was once
RAF Winthorpe near Newark,
Nottinghamshire**

Cycling in London

Now more ubiquitous than the red bus or black taxi, cycling in London is fun and much safer than you'd think. The healthiest way from A to B and a perfect vantage point for sightseeing

above **King Charles Street, Westminster**

opposite **The Minories near the Tower of London**

**The Bicycle as
Ambience Prop**

above **Codebreakers'
bicycles at Bletchley Park
in Buckinghamshire**

opposite **Bicycles for a
pretend gentry, propped
outside Jack Wills' shop,
Burnham Market, Norfolk**

Charly Gaul (1932 – 2005), known as the 'Angel of the Mountains', was a Luxembourgian cyclist who won the 1958 Tour de France with four stage victories. This photograph is from the supplement *opposite* to *Le Miroir* cycling magazine for 20 July 1958. Note the *Le Parisien* Peugot support vehicle.

right A packet of Apisérum Royal Jelly ampoules from an advert on the back page

LE MIROIR DU TOUR 1958

MIROIR-SPRINT
Le Reflet du Sport

SUPPLÉMENT AU N° 632   GEMINIANI HÉROS MALHEUREUX   64 PAGES — 125 FR.

Painted Bicycles 3

above **Orange bicycle by Tatiana Rodriguez, probably about to be festooned with flowers**

opposite **Blue bicycle by Bob Oyoo. A wonderful and eccentric example of a big Dutch bicycle, in wonderful and eccentric Amsterdam**

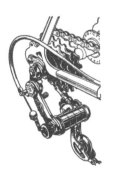

Eric Ellam's cycle repair shop
in Ashby-de-la-Zouch, c1990.
An aromatic shrine to
bicycle mechanicals filled
with wheels and inner tubes,
gear oil and brake blocks

opposite **George Halls
Cycle Centre workshop,
Market Harborough**

Redundant Bicycles

above **Still serving as a useful place to keep the watering can. Marle Place Gardens, Brenchley, Kent**

opposite **Slowly being subsumed and camouflaged in Hartley Wood, Kent**

My boy's bicycle, always left
where it can be tripped over
on dark nights

Atomic Cyclist
Richard Gregory (1996)

'…people who spent most
of their natural lives riding
iron bicycles over the rocky
roadsteads of this parish
get their personalities mixed
up with the personalities of
their bicycle as a result of the
interchanging of the atoms of
each of them and you would
be surprised at the number of
people in these parts who are
nearly half people and half
bicycles…'

The Third Policeman
Flann O'Brien

(Image © 1996 Richard Gregory)

Acknowledgements

My particular thanks must go to Nick Wright, whose help and encouragement have been invaluable

Wilfred Ashley, Bamforth & Co., Bibendum Restaurant at Michelin House, Matt Clark, Christopher Clarke, Dorset County Museum, Richard Gregory, George Halls Cycle Centre, Neil Holman, Stuart Kendall, Andy McKinnon, New York Public Library, Pashley Cycles, Simon Palmer, James Smith and all his team at ACC, David Stanhope, Unsplash (for 'Painted Bicycles 3'), Ken & Hazel Wallace, Philip Wilkinson, Chloe Williams

frontispiece **The bicycle as walking aid, Senate House Passage, Cambridge** *dedication* **Bicycle and beach hut, Harwich, Essex**

© Peter Ashley 2020

ISBN: 9781788840941

The right of Peter Ashley to be indentified
as author of this work has been asserted
by him in accordance with the Copyright,
Designs and Patents Act 1988.

All rights reserved. No part of this
publication may be reproduced, stored
in a retrieval system, or transmitted in
any form or by any means electronic,
mechanical, photocopying, recording or
otherwise, without the prior permission of
the publisher.

British Library Cataloguing-in-Publication
Data. A catalogue for this book is
available from the British Library.

The author and publisher gratefully
acknowledge the permission granted to
reproduce the copyright material in this
book. Every effort has been made to trace
copyright holders and to obtain
permission for the use of copyright
material. The publisher apologises for
any errors or omissions and would be
grateful if notified of any corrections that
should be incorporated in future reprints
or editions of this book.

Book designed by Peter Ashley.

MIX
Paper from
responsible sources
FSC® C124385
FSC
www.fsc.org

Printed in China for ACC Art Books Ltd.,
Woodbridge, Suffolk IP12 4SD, UK

www.accartbooks.com

ACC
ART
BOOKS